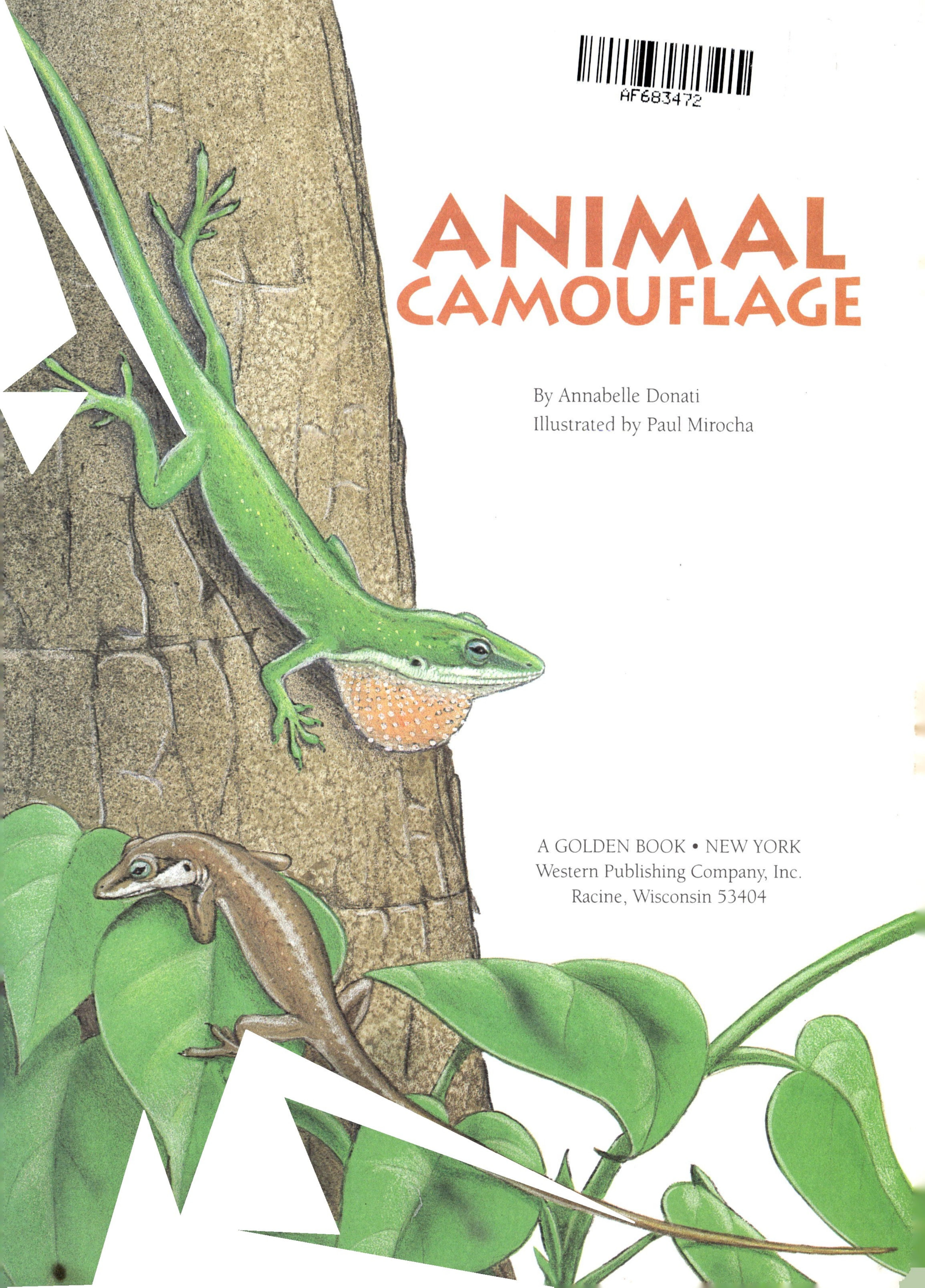

ANIMAL CAMOUFLAGE

By Annabelle Donati
Illustrated by Paul Mirocha

A GOLDEN BOOK • NEW YORK
Western Publishing Company, Inc.
Racine, Wisconsin 53404

Produced by Graymont Enterprises, Inc., Norfolk, Connecticut
Producer: Ruth Lerner Perle
Design: Michele Italiano-Perla
Editorial consultant: Cathy Carlstead, National Zoological Park, Washington, D.C.

 Library of Congress Catalog Card Number: 93-80537 ISBN: 0-307-11328-0/ISBN: 0-307-61328-3 (lib. bdg.) MCMXCVI

Contents

Are walking sticks made of wood?

The walking stick is not made of wood. But when this amazing insect holds its legs against its body, it looks exactly like the branches on which it rests.

The ability of the walking stick and some other animals to blend in closely with their surroundings is called *camouflage*. Camouflage helps animals to survive in two ways. It helps them hide from predators that want to eat them. And it helps them to sneak up on prey that they want to catch.

Tell Me More

When a walking stick is knocked off a plant, it remains perfectly still. To any animal passing by, the insect looks like a broken-off twig.

Amazing but TRUE

The walking stick is the longest insect on earth. It can grow to the astounding length of thirteen inches. That's longer than a page in this book.

Trick sticks

All stick insects, which include the walking stick, the green Indian stick, and the leaf stick, are the same shapes and colors as the tropical twigs and leaves they eat.

Egg drops

Most insects deposit their eggs through tubes called *ovipositors*. Stick insects, however, drop their eggs to the ground from tree branches. The eggs look just like plant seeds, so they, too, are protected by camouflage. When the eggs hatch, the young prepare to climb up the tree trunks. But they wait until night, when they won't be seen.

Why do tigers have stripes?

So they can stalk prey without being noticed! The stripes serve to break up the tiger's shape, making its true outline hard to see in the tall grasses of India and Asia. This type of camouflage is called *disruptive coloration*.

Unlike the cheetah or the lion, the tiger does not try to outrun its prey. To catch its dinner, it must depend on the element of surprise. That's where the tiger's special camouflage helps. The tiger creeps up on its prey and stands as close to it as possible without attracting attention. Then it sprints forward and pounces. But even with the advantage of camouflage, the tiger rarely catches a meal on its first try.

Spots and stripes

Of all the big jungle cats, the tiger is the only one that has stripes. Leopards, ocelots, jaguars, and other wild felines have patterns of spots or patches to break up their outline.

Long and strong

The tiger is the largest and strongest of all the cats. The adult is ten to thirteen feet long from its head to the tip of its tail. It can carry prey more than twice its size.

Amazing but TRUE

All tigers have stripes, but no two tigers have the same pattern of stripes.

Finicky cat

When a tiger has caught a deer, buffalo, or wild pig, it carefully removes the intestines before it starts to eat. If any meat is left over, the tiger hides the uneaten parts in the grass. If it can't get meat, the tiger—a strong swimmer—jumps into the river to catch a fish or a frog.

Meow!

Like other cats—wild or domestic—tigers can purr and meow, and pull in their sharp claws.

Eye contact

Tigers have a pattern on the back of their ears that looks like staring eyes. When facing an enemy, the tiger swivels its ears around. These false eyes make the animal look even more threatening.

Why do crocodiles float like logs?

The crocodile's long, narrow body is covered by rough brownish scales. Because of its shape, color, and ability to remain still for hours, this reptile often looks like a log floating in a muddy jungle river. The two "bumps on the log" are the animal's large bulging eyes, which search for prey above the water while the lower half of its body is hidden underneath.

Look out below!

The crocodile may appear to be a log floating down the river, or a tree fallen onto the riverbank. But any animal that mistakes the crocodile for a harmless log is in big trouble. If a bird, turtle, or small mammal gets too close, the crocodile does not hesitate to snap it up in its strong jaws, drown it, then devour it.

Crock-a-bye babies

The female crocodile digs a hole near the riverbank, lays her eggs in it, and covers them with sand. When the babies are ready to hatch, they start to cry inside their eggs. The mother hears them, digs into the sand, and helps her babies break out of the eggs.

What can you see through sea wasps?

Like all jellyfish, the sea wasp is practically all water. Since the creature is almost transparent, the water, rocks, and plants of the tropical ocean shine right through its bell-shaped body. This makes the animal very hard to see.

Treacherous tentacles

Jellyfish drift gracefully along with the current. Their almost invisible tentacles are spread out, ready to attack any unsuspecting fish, shrimp, or crab that swims by. Some jellyfish use their tentacles like ropes to snare their prey. Others, like the sea wasp, stun or kill their prey with a deadly poison sting.

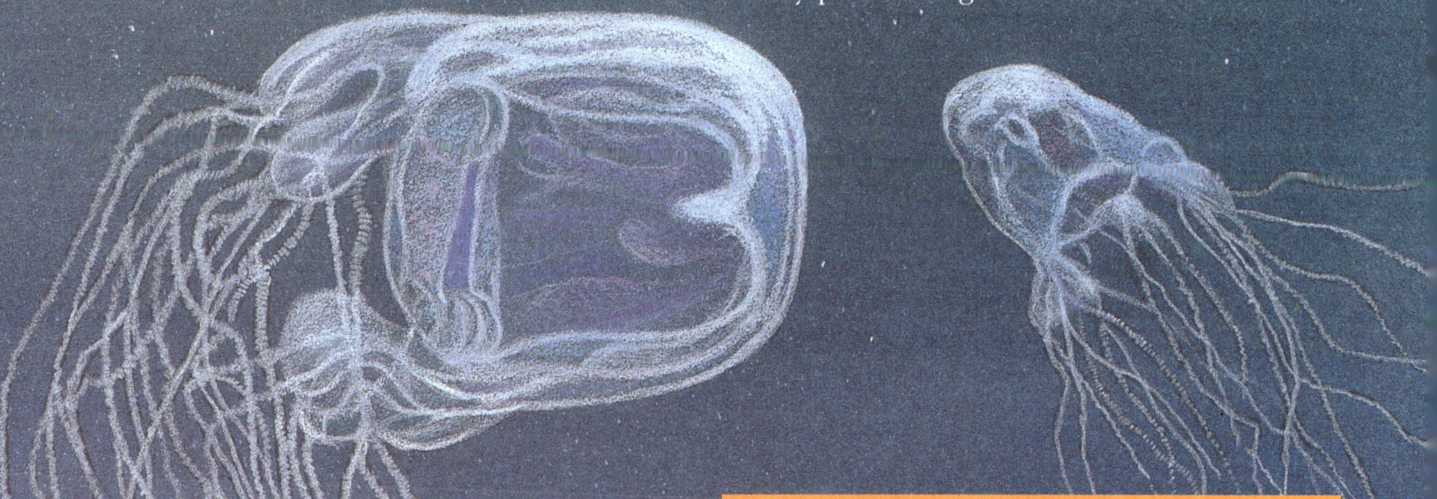

Mirror, mirror

Many fish, such as minnows and sardines, have transparent scales that give their body a silvery sheen. The silvery surface acts like a mirror, reflecting the fish's surroundings. This makes the fish very hard to see among the seaweed, stones, and water that they swim in.

Tell Me More

Jellyfish that have many tentacles are called *medusas*. They are named after the mythical Greek monster Medusa, who had snakes instead of hair growing out of her head.

Why do polar bears hold their nose?

Because they want to hide!

The creamy white color of the polar bear's fur helps it blend in with the snow that always covers the polar region. The only things that could give this big animal away are its shiny black nose and mouth. That's why, some scientists believe, the bear will often cover its muzzle with its white paw, especially when it lies low waiting for a seal to emerge through a breathing hole in the polar ice.

When is a brown bear a white bear?

When it's a polar bear! Although its fur is always a creamy white color, the polar bear belongs to the brown bear family, and it shares many characteristics with its brown-furred cousins. Unlike other brown bears, however, polar bears have fur on the soles of their feet. This helps keep them warm and prevents the bears from slipping on the ice.

Polar bears are good swimmers and spend a lot of time in the water. They use their front feet, which are webbed, for paddling. Their back feet are used for steering.

Tell Me More

Polar bears, the largest meat-eating animals on earth, love to eat seals—especially seal fat. A polar bear can smell a seal several miles away.

Do polar bears ever get frostbite?

Polar bears are well equipped for living on the polar ice cap. They have a thick layer of fat under their skin that works like an extra sweater to keep them warm. Their dense fur also helps them to retain body heat. In addition, their hairs are hollow. This allows the heat from the sun to pass directly to the bear's skin and keeps its fur from freezing.

White and out of sight

The snowy owl is one of the few birds that can survive in the arctic climate. The male's white feathers help it hide in its snowy home. To keep her eggs from freezing, the spotted female lays her eggs among rocks or in the ground. Her spotted feathers help to camouflage her as she sits on her eggs.

Mama bear

Mother polar bears give birth to their cubs inside a deep hole, or *den*. The den is warm and cozy and keeps out the arctic wind. The mother nurses the cubs and never leaves the den, even to eat. In the spring the mother is thin and tired, but the babies are well fed and playful, sliding on the icy hills and exploring their new world.

What makes puff adders puff themselves up?

To frighten its enemies, the puff adder breathes in air so that its body puffs up to twice its normal size. Then, to make itself appear even more menacing, it hisses loudly.

This extremely venomous snake is almost invisible on its home turf in tropical Africa. Although the puff adder is big—three to five feet in length—its skin texture and colors keep it well camouflaged. Other animals could be grazing right on top of the snake without realizing the danger.

Divine cousin

The puff adder's cousin, the vine snake, makes its home in the rainforest. Because this snake lives in trees, it has a green body and a strong grasping tail that coils around branches, looking just like a rainforest vine.

How dangerous is the puff adder?

If a rat or a frog comes close enough, the adder bites its prey with its hollow, pointed teeth, called *fangs*, and injects its venom. The puff adder's fangs are among the longest of any snake's, and its venom is so deadly that it can paralyze and kill a small animal in seconds.

Tell Me More

Unlike most snakes, which always slither along the ground in S-shaped curves, the puff adder can also move along the ground in a straight line.

How many coats do arctic foxes have?

Only one, really. But the arctic fox, which lives near the North Pole, changes the color of its coat with the season. From November until about April, when the polar region is totally covered with ice and snow, many arctic foxes are creamy white. As the snow melts and patches of ground appear, these foxes change their color to brownish gray. Then, around September or October—when the snow starts to fall—their fur turns white again.

Pump it up

The arctic fox's fur is thick enough to keep the animal warm most of the time, but not when temperatures reach forty degrees below zero Fahrenheit. When it gets that cold, the fox's metabolism speeds up, which means its body burns food faster. This raises the animal's body temperature.

Storekeepers

Finding food in the harsh climate of the far north is a year-round struggle for the arctic fox. The most difficult time to find food is during the winter, when most prey have migrated to warmer climates. This is why arctic foxes prepare themselves ahead of time. In summer, when food is more readily available, they capture birds and small rodents, such as lemmings, and collect eggs. They store this food in special hiding places, where it will remain until winter.

Tell Me More

Female foxes, or *vixens*, give birth to their babies in the spring. The babies have a brown coat for camouflage. While the vixen is nursing her young, the male supplies her with food. Sometimes he brings home as many as thirty or forty lemmings a day.

Why are tropical fish so flashy?

Most fish that live in the cold, dark oceans of the world tend to be dull and mottled. But fish that live around the brightly colored coral reefs near the equator are usually a kaleidoscope of shimmering colors and patterns. Many of these tropical fish look more like paintings than living creatures. Yet they almost disappear against the brilliant pinks, blues, violets, and yellows of the surrounding coral and sponge formations.

How do stripes and spots protect tropical fish?

The stripes and spots on fish such as sweetlips and coral rabbitfish break up the fish's outline. This makes it hard to detect among the tropical plants in the shimmering water.

Tell Me More

Some fish travel in large groups called *schools*. When a school of fish, such as zebrafish, travels together, the jumble of stripes or other patterns makes it hard for a predator to single out just one fish to catch.

Amazing *but* TRUE

Some tropical fish, such as butterfly fish, have markings that look just like eyes. It's hard to tell which end of the fish is its head! These false eyes often confuse a predator long enough for the fish to make its escape.

Why do some birds play hide-and-seek?

Most birds fly away when they sense danger. But some birds prefer to stay on the ground. Others will not take off when they are tending their young. When birds don't want to use their wings to escape, their ability to hide—camouflage—becomes their best defense.

Sleek beak

When it holds its long beak straight up, the graceful bittern can hide among the tall reeds of its watery habitat.

Dressed for the season

The ptarmigan's plumage, which is pure white in winter, is a perfect disguise when snow covers the mountains where it lives.

But in summer, after the snow melts, the ptarmigan sheds its feathers and grows a new dress of grayish brown.

Stone clone

A speckled plover becomes all but invisible among the stones and gravel of its seashore home.

Why are some birds' eggs decorated?

Birds build their nests in a variety of places—in trees, on cliffs, and on the ground. Eggs that are not likely to be seen, such as those laid high on cliffs or inside burrows, are usually white. Eggs that are in danger of being seen and eaten by an enemy often have spots or patterns that help them blend with their surroundings.

Leafy

Grouse eggs, which are laid on the ground, have shells that look as if leaves or sticks are painted on them.

Rocky

Tern eggs are almost impossible to see among the pebbles and gravel on which they are laid.

Glossy white

Woodpecker eggs are shiny white. They have no need for camouflage patterns because they are laid inside tree holes.

Twiggy

Gull eggs have a brown speckled shell that hides them among the twigs and rocks on the ground where they are laid.

Amazing but TRUE

Only the outside of the eggshell is decorated. Once their chicks hatch, bird parents quickly remove the broken shells so that the white insides won't attract any predators.

Do pipefish smoke?

Pipefish do *not* smoke. The pipefish gets its name from the shape of its long, narrow, cylindrical body.

A school of pipefish, swaying in the water, looks exactly like a field of underwater plants. The long, thin shape and waving motion of these unusual fish make them all but impossible to distinguish from the sea grasses that grow in the shallow waters where they live.

Tails up

The pipefish has a special tail it can curl around seaweed to anchor itself. Attached this way, the fish can float for hours without moving a muscle.

Papa bearer

Pipefish perform beautiful mating dances, during which they wind their bodies gracefully around one another. The female deposits her eggs in the male's *brood pouch*, a groove on the underside of his body covered by two flaps of skin. The male then fertilizes the eggs and carries them in his pouch until they hatch and the young fish, called *fry*, swim out.

Tell Me More

The pipefish, like its cousin the seahorse, has a body covered with bony plates instead of scales.

Ghosts!

The ghost pipefish has a wider, less streamlined body. It looks like a piece of floating seaweed.

How do deer hide their shadow?

The white-tailed deer's best defense against hungry predators is its ability to hide. Its tawny coat makes this graceful animal practically invisible in its leafy, sun-filtered, woodland environment. But there are more ways for the deer to hide than this.

One way predators find their prey is by looking for telltale shadows that animals cast on the ground. But the deer can reduce the size of its shadow. The animal crouches as low as possible and holds its head and neck in line with the ground. When danger is near, the deer stays perfectly still in this position. Any movement of foliage or sound of hooves in the underbrush would give the animal away.

There's even more. Because of the deer's white belly, predators find it hard to make out the deer's shape in the glare of the sun.

Tell Me More

Most male deer, called *bucks* or *stags*, have antlers growing on top of their head. These antlers look like the branches of a tree. The deer shed their antlers each year in the winter, and new, even larger ones start growing back in the early spring. The delicate new antlers have a soft, fuzzy covering called *velvet*, which is shed in the fall. A stag's antlers are among the fastest-growing structures in the animal kingdom. A healthy stag's antlers can grow as much as four inches a week!

Sun spots

Baby deer, called *fawns*, have a spotted back and shoulders that look like dappled sunlight shining through the trees. These spots are an added measure of camouflage for the young, who are weak and can't run from danger as fast as the adults. As the fawns grow bigger and stronger, the spots disappear.

Hide-and-seek

The female deer, called a *doe*, hides her newborn babies in the woods and comes back to feed and groom them two or three times a day. After each feeding, the doe changes the location of the hiding place to be sure that no predator can sniff out her family.

Why are most tree frogs green?

If they weren't green, tree frogs would have a difficult time hiding from birds, monkeys, and other predators that eat frogs. Their color makes these little jumpers almost invisible among the green leafy tree branches where they—and their predators—live.

Amazing but TRUE

Some tree frogs are poisonous. These frogs, which are usually *not* green, have bright colors and patterns. Instead of hiding from predators, they warn them away with their coloration. Their message is clear: *Eating me is definitely dangerous to your health!*

Brown beauty

The brown tree frog found in parts of South America is colored and shaped like a dead leaf. This frog can hardly be seen when it is climbing tree trunks or sitting among fallen leaves.

World-class hoppers

Tree frogs are found all over the world. They come in many different sizes and shapes. Some do unusual things:

- The bright green squirrel tree frog of North America is tiny. It is only one and a half inches long—a little longer than a paper clip.
- The Japanese foam-nesting tree frog builds a nest on branches that hang over the water. Shortly after the female lays her jelly-covered eggs, the male uses his hind legs to help her beat the jelly into a foam. The foam then hardens, forming a shell around the eggs. When the eggs start to hatch, the foam softens again, and the baby frogs, called *tadpoles*, slip into the water below.
- Wallace's flying frog, which lives in the rainforests of Southeast Asia, has huge webbed feet that open like parachutes as the frog jumps from branch to branch.
- The red-eyed tree frog of Central America has bright red eyes. Its dark eyelids hide the red color. Some scientists believe that frogs can see through their eyelids.
- White's tree frog of Australia can swallow a snail, including its shell, whole.

What makes emus look so hairy?

You would think that the emu, a bird almost six feet tall, would have a difficult time hiding from predators. Not so. This big Australian bird has long, stringy feathers on its body that look like strands of coarse hair. When an emu feels threatened, it lies down on the ground and tucks its head into its hairlike feathers. As long as it remains perfectly still, the emu looks just like the dense bushes on the Australian plains. An enemy passing by may never know the bird is there.

Tell Me More

The emu has tiny wings and a huge body. It cannot fly. Like the ostrich, though, it has long, strong legs, so it can run very fast.

Amazing but TRUE

When an emu mother lays her eggs, they are dull green. After a while, they become black and shiny. An emu lays a lot of eggs—up to fifteen at a time, each weighing about a pound.

Dear old Dad

The male emu is different from most other male birds. The female lays the eggs, but the male builds the nest and sits on the eggs until they hatch. Since hawks, eagles, and foxes are always trying to snatch emu eggs or chicks, Dad rarely leaves the nest until the young are at least a few months old.

How many heads do molochs have?

Although the moloch (MOLL-ick) looks like a dangerous bundle of spines, spikes, and stakes, it is really quite a peaceful animal. When this amazing lizard wants to scare off another animal, it just bends its head down, making its fat hump look like a second head. That confuses the predator, proving that two heads *are* better than one!

Hot, hot, hot

The moloch has an unusual way of collecting water in its hot, dry desert home. Grooves between its scales collect moisture from the air and route it to the moloch's mouth.

Tell Me More

When the moloch is hungry, it sits down near an anthill and licks up the ants as they march in or out of their nest.

Color magic

The moloch, which lives in the Australian desert, is a skillful camouflage artist. It can quickly change its sandy color to red or brown to match its surroundings, making it seem to disappear before a predator's eyes.

What turns anoles gray?

The answer isn't old age! The anole lizard, like the moloch, has the amazing ability to change the color of its skin in just minutes. It can change from pale gray to green to very dark brown. Although its colors often match its surroundings, the anole's change of color—like a chameleon's—is more to show how it feels than for purposes of camouflage. When an anole—young or old—is dark gray, chances are it is sick. When it is a brownish color, it is resting or sleeping. And when it is green or brown, it is feeling fine.

The anole's real means of camouflage is its shape. This lizard's long, thin body looks just like the narrow branches and vines that grow where it lives in the southern United States and the Caribbean.

Tell Me More

The male anole has a colorful throat fan that he displays to attract a female or to show off his strength when he feels threatened. Anoles are quick to bite and often fight each other fiercely when they are competing for territory.

Do dolphins wear rubber wet suits?

Dolphins are marine mammals, not fish, so they don't have scales. Their skin is smooth, soft, and rubbery—like a diver's wet suit. Like many fish, however, dolphins are light-colored on the underside and dark on top. This kind of coloration is called *countershading*, and it helps to protect the dolphins from predators. For example, when a shark is looking down at the dolphin from above, it sees the dark top side of the dolphin, which fades into the deep, dark water below it. But when the shark is below the dolphin, looking up, it sees the dolphin's light side, which is lost in the reflections of sunlit water near the surface.

Wispy!

Like all mammals, dolphins have hair—tiny, nearly invisible wisps on different parts of their body.

Gulp!

Dolphins have more than a hundred small sharp teeth, but these are used only for grabbing and holding on to squid and fish, not for chewing. In fact, dolphins like to swallow their catch whole.

Something fishy

When the steelhead trout hatches in freshwater rivers, its coloration resembles the gravelly stream it is in.

When it moves to ocean water, a few years later, the steelhead becomes countershaded like the dolphin. Its back becomes dark blue and its sides become silvery.

Amazing but TRUE

Smarty!

Dolphins are among the most intelligent animals on earth. They have a great capacity for solving problems. For example, when dolphins are hungry, they leap up out of the

Tell Me More

Though you have come to the last page of this book, you are only beginning to find out about the many ways animals can hide in their environment. Creatures of every description have developed an amazing variety of techniques to conceal themselves from predators and prey.

To start you on your way to new discoveries, here are some more amazing-but-true facts about how animals camouflage themselves.

- The leaf fish of the Amazon looks so much like a leaf that it even has lines that look like the leaf's veins.
- When a herd of zebras travels together, the animals' black and white markings make it hard to tell where one zebra ends and the next begins. This makes it hard for a predator to single one out for capture.
- The sponge crab carries live sponges on its back. The sponges conceal the crab's body and even make it look like a sponge itself.
- The flower mantis looks just like a flower as it perches on a stem, waiting for its next meal—a butterfly—to land.
- The cuttlefish is a squid that can change its color in less than a second!
- The shape and color of the thornbug's body mimics the thorns on roses and other thorny plants. This makes the bug invisible when sitting on the stems.
- The flight movements of the oriental leaf butterfly are just like those of a leaf floating in the breeze.
- The wobbegong shark has a fringe of wiggly growths on its head. It looks just like a rock covered with seaweed.

INDEX